The Culinary Traveler: Volume 17
by The Editors of The Evergreen Centre and Chart Atlas
ISBN: 9781989647820
First published November 2025
Toronto, Ontario

Publisher: The Evergreen Centre

Publisher's Cataloging-in-Publication Data
The Culinary Traveler
The Culinary Traveler: Volume 17 / The Editors of The Evergreen Centre and Chart Atlas.

Summary: This volume, focused on the Blueprint for Margin Mastery, provides a mandate for driving maximum cash flow in gastronomy by merging culinary prestige with unmatched cost efficiency. It delivers high-impact, scalable solutions where technique and concept, not expensive ingredients, create the "WOW" factor and high profit. Key strategies include utilizing low-cost staples like caramelized onions and leek ash for high-margin, high-craft dishes, emphasizing Hyper-Margin, Theatrical Profit, and a Michelin Economy approach to de-risk modern restaurant operations.

Identifiers: 9781989647820

Subjects: Culinary Arts. | Restaurant Management. | Cost Control. | Gastronomy. | Food Service Management.

Classification: 647.95–dc23

The Culinary Traveler Vol.

EDITOR VOLUME 17 INTRO

17

HIGH WOW HIGH PROFIT

Hyper-Margin: Use negligible-cost leek ash for a truffle-like effect.

Theatrical Profit: 92% margin on "Fizz Phantom Cubes" (low-cost, high-impact beverage showpiece).

Defiant Flavor: Leverage low-cost staples (caramelized onions) and temperature contrast for high-WOW desserts.

De-Risking: Less than 4% Costs on Haute Cuisine Main

Michelin Economy: Turning Pantry Staples into high-wow exclusive ingredients that drive word-of-mouth and loyalty

Flex: Charting heretofore uncharted flavors.

MISSION:

Drive maximum cash flow in gastronomy by merging culinary prestige with unmatched cost efficiency.

THE MANDATE (Buzz & Money):

Deliver high-impact, scalable solutions where technique and concept, not expensive ingredients, create the WOW factor.

The Welsh Leek Ash & Black Garlic Umami Broth
OVERVIEW + INGREDIENTS

The Welsh Leek Ash & Black Garlic Umami Broth is not just a liquid—it's the concentrated essence of the New British Pantry. This jet-black, savory elixir is the core of our **UK Heritage Luxe Kit**, delivering intense, mature flavor without expensive base ingredients.

We elevate humble leek ash and proprietary black garlic into a high-impact, high-margin culinary statement.

A single, rich teaspoon transforms everything from stews to vinaigrettes, embodying the fusion of ancient technique and modern profitability. This deep, haunting broth is your new secret weapon.

Ingredients	Measurement (Batch)	COGS Strategy
Filtered Water	4 Litres	Low-cost, bulk base liquid.
Malt Vinegar (Yorkshire)	500 ml	Pantry Item: Acidity and preservative.
Black Treacle	400g	Pantry Item: Color, sweetness, and viscosity.
Marmite/Yeast Extract	600g	Core Umami: Cost-efficient, bulk flavor base.
Fine Sea Salt	200g	Standard seasoning and preservative.
Welsh Black Garlic	150g	WOW Factor: High-impact flavor driver; used sparingly for effect. House-made substitution follows
Leek Ash Powder	10g	Hyper-Margin: Charred bulk veg trimmings; negligible cost per jar.
Xanthan Gum	10g	COGS Strategy: A low-cost thickener for desirable viscosity.

The Welsh Leek Ash & Black Garlic Umami Broth
METHOD + BUSINESS CASE

Follow these steps precisely to create a stable, concentrated, jet-black savory broth.

Prepare the Leek Ash: Slowly char the white and green parts of bulk leeks until they are completely black and brittle. Ensure all moisture is driven out. Grind this material into a very fine, soot-like powder. Store in an airtight container immediately. (This is the hyper-margin ingredient.)

Dissolve Solids: In a large, non-reactive pot, combine the Filtered Water, Malt Vinegar, and Black Treacle. Heat gently and stir continuously until the treacle is fully dissolved and incorporated.

Add Umami Base: Stir in the Marmite/Yeast Extract and the Fine Sea Salt until they are fully dissolved. Bring the entire mixture to a gentle simmer.

Infuse Black Garlic: Add the whole cloves of Welsh Black Garlic. Reduce the heat and gently simmer the broth for 30 minutes to allow the complex, sweet-savory black garlic flavor to permeate the liquid fully.

Strain Base: Remove the pot from the heat. Strain the liquid through a fine sieve to remove the garlic solids. Discard the solids (or reserve for staff meals).

Thicken and Color: Allow the broth to cool slightly - to about 60-70°C (140-160°F) . Using an immersion blender on high speed, slowly incorporate the Xanthan Gum (add gradually to avoid lumps). Once smooth and slightly thickened, blend in the Leek Ash Powder until the broth achieves a uniform, deep, opaque black color.

Bottle and Seal: Pour the hot broth into sterilized jars or bottles and seal immediately to preserve its stability and flavor profile.

WOW FACTOR:

- **Jet-Black & Mysterious:** Visually stunning, unique.
- **Leek Ash + Black Garlic:** Innovative, gourmet, surprising.
- **"Secret Weapon":** Potent, versatile, chef-level impact.
- **"Luxe Kit":** Premium, elevated British cuisine.

PROFIT MARGIN:

- **Hyper-Margin Leek Ash:** Near-zero cost, high-value ingredient (waste to gourmet).
- **Low-Cost Bulk Bases:** Water, vinegar, treacle, Marmite are cheap, effective fillers.
- **Black Garlic (Controlled Use):** High-impact flavor, used sparingly for max effect.
- **Concentrated Product:** High perceived value, justifies premium price, low consumer usage per serving.

Bottom Line:

Visually arresting, uniquely flavored, and built on an incredibly efficient cost structure for high profitability.

The Welsh Leek Ash & Black Garlic Umami Broth
WELSH BLACK GARLIC HACK

PRO-TIP: Black Garlic Production (The Utility Hack)

Authentic Welsh Black Garlic is a luxury component, but your kitchen can produce a high-quality substitute using standard equipment. The process requires a consistent, controlled environment: low heat (60 C to 70 C) and high humidity for 3–4 weeks.

The Practical Method: Use an inexpensive rice cooker or slow cooker set permanently to the 'Keep Warm' setting.

Prep: Wrap whole garlic bulbs tightly in foil and place them inside the cooker.

Seal: Take a large sheet of heavy-duty aluminum foil and cover the entire top opening of the cooker, wrapping it down around the exterior sides.

Secure: Place the cooker's normal lid on top, then use high-temperature aluminum foil tape (often used for sealing kitchen ventilation or oven components) to securely seal the junction between the lid and the outer casing. This traps essential moisture, preventing the garlic from drying out.

Monitor: Check a single bulb after 2–3 weeks. The cloves are ready when they are soft, black, and chewy (Maillard reaction complete).

Best Garlic for Black Garlic (Restaurant):

Hardneck Garlic (Large Cloves)

Why: Larger cloves mean less peeling time (efficiency!), and they often yield a richer, more complex umami flavor.

Look For: Varieties like 'Music' or 'Spanish Roja'.

Avoid: Small-cloved softneck varieties (too much tedious peeling).

The UK Heritage Luxe Kit:
Your Margin Accelerator + New Classic Trio

In today's competitive landscape, true distinction lies in the integrity of your flavor story. The UK Heritage Luxe Kit is not a collection of recipes; it is the strategic, aspirational tool designed to integrate culinary heritage with modern efficiency, allowing your team to achieve peak creative control and unprecedented quality.

This kit empowers your kitchen to unlock a new level of menu prestige while ensuring operational grace. In addition to the **The Welsh Leek Ash & Black Garlic Umami Broth,** our number 1 recipe of the Trio, we've engineered two two other foundational ingredients to be your signature flavor anchors.

Culinary Control: A New Standard of Flavor Integrity

The most ambitious menus require complete control over flavor agents. By bringing complex profiles in-house—and replacing inconsistent imported purees and stocks with our high-yield foundations—you gain mastery over your plate. This is the difference between serving a concept and delivering a defining experience.

Recipes Follow

For the complete Profit Business Blueprint Kit, including master recipe files, kitchen cards, scale production methods, cost/margin spreadsheets and marketing and sales assets visit:

- **www.explorechartatlas.com/blueprints**

Completing the Set:

2. The Mermaid's Larder Finishing Oil

- **The Signature Flourish:** This is the High-Prestige Drizzle—the final, vibrant signature on your plate. It carries a bright, aromatic flash of coastal herbs, lending a floral complexity that elevates a simple dish to an elegant experience.

- **The Elegant Solution:** This oil is a triumph of perception. Built on a low-cost rapeseed oil base, its concentrated power means only a tiny amount is needed per plate. You deliver maximum flavor impact while maintaining excellent margins, proving that the most memorable details can also be the most efficiently refined.

3. The Captain's Brown Sauce

- **The Culinary Elevation:** Imagine the rich resonance of deep, syrupy umami, achieved instantly. This glaze elevates a classic by transforming low-cost kelp into a unique oceanic base, delivering a flavor depth usually reserved for expensive stock reductions. It is the secret to a syrupy, malty finish—perfect for glazing heritage meats, finishing root vegetables, or anchoring a savory pie base.

- **The Operational Grace:** Achieving this depth is now effortless. We bypass the labor and cost of multi-day reductions. The shelf-stable base ensures zero compromise on consistency and eliminates the risk of waste, rewarding your artistry with superb efficiency.

The "Mermaid's Larder"
Cotswold Lavender, Thyme & Samphire Herb Oil

Ingredients	Measurement (Batch)	COGS Strategy
High-Oleic Rapeseed Oil	4 Litres	Low-cost bulk base (90%+ of product).
Pickled Samphire (drained)	200g	Provides brine and color.
Cotswold Thyme (Dried)	20g	Concentrated flavor, extremely low cost.
Cornish Sea Salt & Herb Infusion	50g	Seasoning and coastal notes.
Cotswold Lavender Honey	30g	WOW Factor: Aroma and slight stabilizer; low volume used.
Culinary Lavender Flowers (Dried)	10g	Pure floral aroma.
Fresh Parsley	50g	Color Boost: Boosts vibrant green

Method

1. **Samphire Pre-Treatment:** Drain samphire. Blanch quickly in boiling water (30 seconds) then plunge into ice water. Pat completely dry to stabilize color and remove surface moisture. Finely chop.

2. **Herb Paste:** Blend Blanched Samphire, Thyme, Salt, Infusion, and Parsley with ≈50ml Rapeseed Oil until a smooth, vibrant green paste forms.

3. **Oil Infusion:** Combine remaining Rapeseed Oil, Lavender Honey, and Lavender Flowers in a large container. Whisk vigorously until the honey is dispersed into tiny droplets.

4. **Combine & Rest:** Add the herb paste to the oil. Stir thoroughly and rest at room temperature for 24–48 hours to infuse.

5. **Final Filtration:** Slowly strain the oil through a fine-mesh cloth or sieve for a clear finish. (Pro-Tip: For highest clarity and shelf-life, gently heat to 70 C for 1 hour before cooling and straining.)

6. **Bottle:** Pour the finished, bright green oil into sterilized jars/bottles and seal.

Barrel-Aged Malt & Seaweed Sauce
Captain's Brown Sauce

Ingredients	Measurement (Batch)	COGS Strategy
Barrel-Aged Malt Vinegar	2 Litres	Essential flavor base: malty, woody, complex.
Filtered Water	1.5 Litres	Cuts vinegar intensity; aids viscosity.
Pitted Dates	500g	Natural, fruity sweetness, body, and dark color.
Molasses (Black Treacle)	300g	Rich, sticky texture and deep caramel notes.
Dark Brown Sugar	250g	Balances high acidity.
Dried Kelp Powder	10g	Umami Depth Charge: Intense oceanic savory flavor.
Ground Spices (Ginger, Pepper, Allspice, Garlic/Onion Powder)	25g Total	Core "Brown Sauce" complexity and savory foundation.
Cornflour (Corn Starch)	50g	Low-cost thickening agent.
Cold Water (for slurry)	100ml	Slurry base.

Method

1. **Prepare Base:** In a large, non-reactive pot, combine Vinegar, Water, Dates, Molasses, and Sugar.
2. **Simmer & Soften:** Bring to a simmer, stirring until sugar dissolves. Cook gently for 40 minutes to fully soften dates and integrate flavors.
3. **Add Agents:** Remove from heat. Stir in Dried Kelp Powder and Ground Spices. Simmer for an additional 5 minutes.
4. **Blend Smooth:** Carefully transfer and blend on high until perfectly smooth. (No date or kelp chunks should remain).
5. **Thicken:** Return to a low simmer. Whisk Cornflour into Cold Water to form a smooth slurry.
6. **Finish:** While whisking sauce constantly, slowly pour the slurry in. Stir for 1–2 minutes until syrupy, ketchup-like consistency is achieved.
7. **Storage:** Pour piping hot into sterilized, airtight jars/bottles and seal tightly for maximum shelf-life.

The Black Vinegar Empire
Test Kitchen Previews

The Project: Elevating Pantry Staples for Profit

Our goal is simple: transform a common pantry item—**Black Vinegar**—into high-margin, high-impact culinary solutions.

The secret? Concentrated flavor architecture, not complexity.

Black Vinegar forms the core of our "Vinegar Empire Kit." It's a unique ingredient blending the malty richness of the UK with the savory tang of classic Asian ferments. This is East-West heritage engineered for profit.

Our Goal:

$8 Retail Concepts (20 bites) with COGS below £1.20.

We achieve this by harnessing **Black Vinegar's** intense, low-volume flavor to reduce reliance on more expensive primary ingredients, ensuring exceptional margins.

The Umami Depth Charge

Black Vinegar is a sophisticated, deeply aged Chinese condiment. Crafted from glutinous rice, millet, or sorghum, it undergoes a long fermentation process. Its signature flavor is malty, smoky, intensely savory (umami), with a rich, dark color and a subtle sweetness that distinguishes it from harsher vinegars. It provides a final, profound finishing depth in glazes and sauces.

Methodology:

Metric	Target	Result	Strategy
WOW	Factor 10/10	Fusion, texture, theatricality.	Unique East-West concepts.
Target Margin	89% + COGS	<£1.20 for £8 MSRP.	Low-cost bases (sugar, pastry, water).
Labor Tactic	Minimal	Batch-prep in under 2 hours.	Simple methods; focus on assembly/packaging.

Showcase Solution:
Black Vinegar Shrub Ice Cubes
(The "Phantom Fizz Cubes")
Test Kitchen Reveal

The Concept

This is one of our high-wow, high-margin solutions: a frozen ginger-fruit shrub. It's a "phantom elixir" that connects colonial punches with Chinese vinegars, melting into a visually striking ruby tint in the glass.

Essential bar mixer for tonic highballs (e.g., in a gin & shrub fizz) or retailed as "cube drops" in chilled pouches for non-alcoholic pop-ups and summer menus.

Yields: Approx. 200 cubes (10 packs of 20)

Juice Options Explored:

Tart Cherry Juice: Excellent balance of sweet and tart, vibrant ruby color, strong flavor profile.
Cranberry Juice: Readily available, very cost-effective, sharp tartness, good red color.
Elderberry Juice: Strong "foraged" brand narrative, deep dark color, unique earthy-tart notes.
Pomegranate Juice: Complex sweet-tart, slightly tannic, good color.

Ingredients & Flavor Purpose:

- **Core:** Black Vinegar (200ml) provides the malty sour core.

- **Aroma & Balance:** Fresh Ginger (100g, juiced) adds spicy heat, and Fruit Juice Concentrate (150g, e.g., Tart Cherry; Cranberry; Elderberry; Pomegranate) lends a tart veil and natural color. Or, upgrade to Yuzu or Papaya for a premium tier.

- **Sweetener:** Demerara Sugar (100g) sweetens the shrub and adds a caramel undertone.

- **Dilution:** Water (800ml total) for volume and balance.

Preparation Method:

- **Make Shrub Base:** Puree ginger, fruit concentrate, sugar, and black vinegar. Strain into a jug.

- **Dilute:** Stir in water and adjust sugar for a balanced, sharp lemonade-like flavor.

- **Freeze & Serve:** Pour into ice cube or spherical molds and freeze solid. Drop into sparkling water, tonics, or cocktails.

WOW Rating:

10/10 – A "phantom-like melt" with a progressive sour-ginger zing, transforming drinks into a dynamic, evolving elixir. This customizable ritual feels like alchemy in a glass.

Profit Rating:

10/10 - (92% Margin).
Total COGS: £0.91 per 20-cube retail pack.

Due to its sugar and vinegar content, the Black Vinegar Shrub Ice Cube will melt faster than plain ice, rapidly infusing drinks with its complex flavors and delivering its "dynamic elixir" wow factor as intended.

The Black Vinegar Empire Kit:
The 90% Margin Trio Overview

The 90% Margin Trio: Flavor and Finance

These three concepts deliver maximum sensory impact at minimum ingredient cost, creating essential bar snacks and retail products. Each recipe is designed for rapid batch preparation and guaranteed high profit margins.

1. Black Vinegar & Soy Caramel Bites
The "Silk Road Nuggets"

This "Silk Road Nugget" recipe is a game-changer as it masterfully balances the complex smoky tang of black vinegar with soy's umami and honey's sweetness, transforming a simple caramel into a sophisticated, visually striking sweet-sour relic that elevates any tasting menu or cocktail bar offering.

Method: Combine 500ml black vinegar, 300ml soy sauce, 400g honey, and 200g granulated sugar in a heavy-bottomed saucepan. Heat over medium until sugar dissolves, then simmer for 25-30 minutes (or to 118-120°C for soft caramel) until a thick, syrupy consistency is achieved. Pour into prepared molds or a lined tray, sprinkle with 20g sea salt, and refrigerate for 1 hour until firm before cutting into approx. 200 bites. These "Silk Road Nuggets" offer a sweet-sour, umami-rich caramel experience, ideal as an amuse-bouche or cocktail bar " guilt-free chew.

Pro Use: Amuse-bouche, petit four, or cocktail chew.

WOW Rating: 10/10 – Tangy melt reveals surprising saline depth, resulting in an addictive flavor tension. The addictive "sweet-storm" dissonance, where the vinegar's sharp lift erupts through the honey's velvety sweetness, is a masterpiece. Its glossy, jewel-like visuals and historical narrative transform a simple candy into an Instagram-worthy conversation starter.

Financial Snapshot: Retail: £8.00 / COGS: £1.10 (90% Margin)

2. Black Vinegar & Puff Pastry Bites
The "Vinegar Veil Crisps"

This "Vinegar Veil Crisps" recipe is a game-changer by fusing the delicate flakiness of puff pastry with a bold black vinegar glaze and toasted sesame, creating an elegant "veiled crisp" that elevates traditional British textures with an unexpected, sophisticated tang.

Method: Roll thawed puff pastry to desired thickness, cut into squares. Reduce black vinegar by half to a syrupy glaze. Brush pastry squares with neutral oil, bake at 200°C (400°F) for 12-15 minutes until puffed, then immediately brush with warm glaze and sprinkle generously with toasted sesame seeds.

WOW Rating: 10/10 – A shattering buttery crunch meets balsamic tang and explosive nuttiness.

Profit Rating: 10/10 (90% Margin at £1.00 COGS / £10.00 Retail).

---

3. Mini Black Vinegar Gel Cubes for Dumplings
The "Burst Relic Orbs"

This "Burst Relic Orbs" recipe is a game-changer by encapsulating a potent garlic-black vinegar aspic within a delicate dumpling wrapper, offering an interactive "orb burst" of fermented fire and umami that elevates dim sum to a theatrical, multi-sensory experience.

Method: Bloom gelatin in cold water. Gently warm black vinegar, water, and fresh minced garlic (do not boil), then stir in bloomed gelatin until dissolved. Strain into cube molds and refrigerate for 2 hours until firm. Place one gel cube in each thin, round dumpling wrapper, pleat, and seal edges. Steam for 8-10 minutes until translucent.

WOW Rating: 10/10 – An icy snap unleashes concentrated fermented fire, leading to a huge umami build; the interactive "orb burst" is mesmerizing.

Profit Rating: 10/10 (91% Margin at £0.92 COGS / £10.22 Retail).

For the complete Profit Business Blueprint Kit, including all four Black Vinegar Empire master recipe files, kitchen cards, scale production methods, cost/margin spreadsheets and marketing and sales assets visit: **www.explorechartatlas.com/blueprints**

SPANISH CITRUS FIRE
Taste the Uncharted.

Ignite your senses with the fiery spirit of Seville, kissed by the Mediterranean breeze. Our "Spanish Citrus Fire Elixir Oil" is a vibrant, Uncharted sensory journey, capturing the essence of sun-drenched Spanish citrus groves and the smoky soul of La Vera cuisine.

This oil achieves its high WOW factor by delivering a complex flavor paradox: Bright Citrus Fire against Smoky Ember Depth and a Mediterranean Sea Air Brine. This specific combination of blood orange, smoked paprika, chili, and cured olives is groundbreaking and unavailable in the gourmet market.

We meticulously cold-infuse the zest and juice of ripe blood oranges with deeply flavorful Spanish smoked paprika (La Vera Pimentón), Calabrian chili flakes, and the rich, briny note of cured Spanish Manzanilla olives. Grounded in a robust extra virgin olive oil, this elixir is designed to electrify your palate and provide your menu with a unique, high-craft signature.

SPANISH CITRUS FIRE
Production Protocol: Cold Infusion Master File

The quality of this Elixir relies on the meticulous cold-infusion method, which avoids heat degradation and preserves the delicate floral notes of the blood orange.

The Components:

1. **Base Oil:** 1.5 Liters High-Quality Extra Virgin Olive Oil (Spanish Arbequina or Picual recommended).
2. **Citrus:** Blood Orange Zest from 6-8 large Blood Oranges (thinly zested, pith avoided), 150 ml Blood Orange Juice (freshly squeezed, strained).
3. **Spice:** 30 grams Spanish Smoked Paprika (La Vera Pimentón), 10 grams Calabrian Chili Flakes.
4. **Brine:** 200 grams Spanish Green Olives (Manzanilla or Gordal, brine-cured, pitted, gently crushed/bruised).
5. **Aromatic:** 4-7 Lightly Roasted Garlic Cloves (optional, for savory depth).

Cold Infusion Method:

1. Combine the zest, juice, paprika, chili, and bruised olives in the Extra Virgin Olive Oil base.
2. Agitate gently to combine and infuse at room temperature.
3. Infuse for a period of 2 to 3 weeks, tasting the oil regularly to monitor flavor development.
4. Strain to the desired clarity: either a rustic finish with fine olive pieces or a completely clear oil.
5. Store in dark, airtight bottles to maximize shelf life and color stability.

FOOD SAFETY CCP: Due to the inclusion of low-acid ingredients (garlic and olives) in an anaerobic oil environment, this product must be refrigerated immediately after straining and used within two weeks, or frozen for longer storage. This is a cold-infused product, not shelf-stable. For this "Spanish Citrus Fire" elixir to be legally and safely sold as a retail product in the U.S., a process authority review and the implementation of a scheduled process (likely involving validated acidification or stringent cold-chain management with explicit shelf-life limitations) would be a critical requirement.

The "Smoked Orange & Olive" Martini

Recipe:
- 2.5 oz London Dry Gin
- 0.5 oz Dry Vermouth
- 0.25 oz "Spanish Citrus Fire" Elixir (its cured olive notes add unexpected depth!)

Method: Combine ingredients in a mixing glass with ice. Stir 30-40 seconds until well-chilled and diluted. Double-strain into a chilled coupe. Garnish with a single caper berry.

Flavor Profile: This Martini offers a complex, elegant fusion. Juniper is enhanced by the elixir's smoky paprika, bright blood orange, and a surprising, savory depth from its infused cured olives, while the caper berry garnish provides a final, piquant pop, creating a rich, subtly tangy, and utterly unique experience.

SPANISH CITRUS FIRE
Tasting Notes: Seville Vibrancy & Smoky Ember Depth

Initial Aroma
A vibrant burst of sun-drenched Seville orange zest, mingling with the enticing smoky aroma of Spanish paprika. A subtle, briny whisper emerges, reminiscent of Mediterranean sea air, underpinned by a gentle fruity olive oil base.

First Palate
Zesty and bright citrus immediately awakens the palate, with the distinct floral-berry notes of blood orange taking center stage. A clean, refreshing vinegar-like acidity from the citrus juice cuts through the oil's richness.

Mid-Palate
Smoked paprika unfolds its complex, ember-like depth, providing a savory and grounding warmth. The briny notes of cured olives become apparent, offering a savory-salty counterpoint to the citrus sweetness. A gentle, building warmth from Calabrian chili flakes adds a playful spark.

Finish
Lingering citrus brightness intertwines with smoky paprika warmth and a subtle salty-briny finish. The clean olive oil base provides a smooth mouthfeel, while the chili warmth leaves a pleasant, lingering tingle. The overall sensory impression is intensely vibrant and deeply grounded.

Culinary Uses for Spanish Citrus Fire: From Vinaigrette to Uncharted Cocktail

Conventional Uses with "Spanish Fire" Elevation

- **Vibrant Vinaigrettes:** Create zesty, Spanish-inspired vinaigrettes for salads with bitter greens (arugula, endive) or grain salads.

- **Citrus Marinades:** Use as a quick marinade base for grilled chicken, shrimp, or fish, imbuing them with smoky citrus flavor.

- **Finishing Oil:** Drizzle generously over grilled sardines, prawns, or white fish. A sophisticated finishing touch for roasted potatoes, peppers, or eggplant.

- **Tapas Enhancement:** Serve as a flavorful dipping oil for crusty bread, or drizzle over Spanish tapas like patatas bravas or tortilla.

Avant-Garde & "Uncharted" Culinary Explorations

- **Emulsified Citrus-Paprika Sauce:** Create vibrant, light emulsions to accompany refined seafood or vegetable dishes, using the Elixir as the unexpected flavor base.

- **Aioli/Mayonnaise Boost:** Stir into aioli or mayonnaise for a complex "Spanish Fire Aioli" used for dipping or unique sandwich spreads.

- **Dessert Accents:** Drizzle sparingly over citrus-based desserts (orange sorbet, grapefruit granita, olive oil cake) to add a sophisticated, surprising savory-spicy-briny contrast.

- **Cocktail Innovation:** Use a few drops in a gin or vodka martini for a Mediterranean "Uncharted Martini." It's also perfect for rimming a savory margarita glass, infusing it with citrus, olive, and spice.

-

SPANISH CITRUS FIRE
SELLING THE IRREPLACEABLE

In an era where culinary distinction is paramount, restaurants are increasingly seeking unique flavor signatures that can't be sourced off the shelf.

Enter the "Spanish Citrus Fire" elixir, a groundbreaking cold-infused oil transforming dishes and defining the very essence of innovative cuisine.

This isn't just infused oil; it's a meticulously crafted blend of vibrant blood orange (zest and juice), the smoky depth of La Vera Pimentón, the subtle kick of Calabrian chili flakes, and the briny complexity of cured Spanish olives, all steeped gently in high-quality extra virgin olive oil.

Over 2-3 weeks, this cold infusion avoids heat degradation, preserving delicate aromatics and creating an unparalleled flavor profile that's both bold and nuanced.

Why it's a game-changer for the professional kitchen:

Unrivaled Flavor Depth: It delivers a sensory journey—smoky, tangy, sweet, spicy, and umami—that elevates everything from grilled seafood to roasted vegetables, creating a memorable "taste footprint" for the restaurant.

Signature Ingredient Story: Being house-made allows chefs to tell a compelling story, highlighting craftsmanship and bespoke flavor development, enhancing the dining experience and building brand identity.

Culinary Versatility: A finishing drizzle, a marinade base, or a crucial element in a complex sauce, its distinct profile provides endless avenues for creative application.

Exclusivity: This specific, handcrafted fusion is unobtainable elsewhere, giving the restaurant a true competitive edge and solidifying its reputation for adventurous, high-quality, and unique culinary offerings.

This "Spanish Citrus Fire" is more than an ingredient; it's an edible emblem of a restaurant's commitment to pushing flavor boundaries, turning every plate into an "Uncharted" experience.

Endive & Spinach Linguine with Citrus-Almond Pesto & Golden Raisins
THE HARMONY OF *TERROIR* -PERFECTING THE FAMILIAR

In a culinary landscape defined by audacious flavors, there is profound artistry in perfecting the familiar. This dish is a testament to the Uncharted philosophy: an approachable main course of understated elegance engineered for immediate WOW factor and impressive profitability.

It marries the delicate, fresh bitterness of endive (beautifully mellowed by spinach) with a bright, aromatic citrus-almond pesto. The unexpected sweetness of golden raisins provides a luxurious counterpoint, balancing the greens and delivering a complex, yet surprisingly comforting, flavor profile.

This recipe transcends the ordinary, transforming a beloved classic into a sophisticated experience that confirms the essence of high-craft gastronomy lies in precision, balance, and the ability to optimize margins through intelligent sourcing.

Endive & Spinach Linguine with Citrus-Almond Pesto & Golden Raisins
THE METHOD

Ingredients

For the Pesto:

- 3 cups (packed) fresh baby spinach
- 1 large head Belgian endive, thick core removed, roughly chopped
- ½ cup golden raisins
- ½ cup toasted slivered almonds (or pine nuts)
- 2 cloves garlic, roughly chopped (or 1 clove roasted)
- ½ cup grated Pecorino Romano cheese
  ½ cup extra virgin olive oil, plus more for drizzling
- 2 Tbsp. fresh citrus juice, plus zest of ½ citrus fruit (see Refinements)
- Salt, freshly ground black pepper to taste
- Optional: Pinch of red pepper flakes

For the Linguine & Serving:

- 1 pound linguine pasta (bronze-die recommended)
- Salt, for pasta water
- Reserved pasta water (about 1 cup)
- Extra virgin olive oil, for serving
- Freshly grated Pecorino Romano, for serving
- Optional Garnishes: Toasted slivered almonds, extra golden raisins, fine strips of fresh endive, matching citrus zest.

Method

1. **Blanch Greens:** Briefly blanch chopped endive (1 min) and spinach (30 sec) in salted boiling water. Immediately plunge into an ice bath, drain well, and squeeze out all excess water—crucial for pesto texture.
2. **Toast Almonds:** Toast almonds in a dry skillet until fragrant. Let cool slightly.
3. **Process Pesto Base:** In a food processor, combine blanched greens, raisins, almonds, garlic, and Pecorino Romano. Pulse until roughly chopped.
4. **Emulsify:** With the processor running, slowly drizzle in the olive oil until the pesto is slightly coarse but creamy.
5. **Flavor & Season:** Add the chosen citrus juice and zest. Season generously with salt and pepper.
6. **Cook Linguine:** Cook pasta until al dente, reserving ~1 cup of the starchy water.
7. **Combine & Serve:** Drain the linguine and immediately add it to a large bowl with the pesto. Toss thoroughly. Gradually add the reserved pasta water, a tablespoon at a time, until the sauce emulsifies to a glossy, creamy texture.
8. **Garnish:** Serve immediately. Drizzle with EVOO and garnish with endive leaves, Pecorino and chosen toppings.

Endive & Spinach Linguine with Citrus-Almond Pesto & Golden Raisins
BUSINESS, MARKETING , SALES + TRAINING

The Strategic Edge: Linguine with Endive & Spinach Pesto

WOW Factor: 10/10 (Unexpected Combo & Finesse)
PROFIT Factor: 10/10 (~90% Margin)

A menu asset engineered for maximum guest appeal and unmatched profitability. Culinary excellence meets intelligent cost control.

The WOW Factor: Sophistication Sells

Vibrant, electric-green pesto with golden raisins – instantly photogenic. We elevate familiar pasta with a sophisticated dance: delicate blanched endive bitterness meets sweet golden raisins and bright citrus (e.g., Meyer Lemon). Complex yet approachable, this memorable profile drives repeat orders. Glossy, perfectly emulsified sauce clinging to bronze-die linguine ensures a professional finish.

The PROFIT Factor: Margin Mastery

Low-cost, high-impact greens (spinach, endive) form the volume base, allowing aggressive pricing. Pesto batch-preps, cutting labor. Toasted slivered almonds offer cost-effective luxury. Efficient green utilization minimizes waste.

FOH Training: The Story Hook

"Refined Italian elegance—a study in harmonious contrast. Humble, quality pasta, precise technique, surprising depth." Highlight endive bitterness, sweet raisins, bright citrus. Emphasize the sauce's professional gloss. (Dietary: Nuts, Dairy, Gluten.)

Marketing & Culinary Notes: Refinement & Strategy

Citrus: Lemon (sharp), Meyer Lemon (sweet/floral), Blood Orange (vibrant/bold).
Craft: Inner endive leaves, quality toasted almonds. "Rustic yet refined" pesto. Starch-rich pasta water for emulsification. Elegant nests, subtle garnish.

INTERMEZZO PIVOT: Endive & Spinach Granita

WOW Factor: 10/10
(Innovative & Unexpected)

PROFIT Factor: 10/10
(~90% Margin)

A daring frozen asset
for an avant-garde
menu. Innovation meets
intelligent cost control.

WOW Factor: Visually striking electric-green ice. Leads conversations.
PROFIT Factor: Low-cost greens, batch-prep friendly, high margin
Sales & Marketing (Core Message)
Story Hook: "A daring frozen expression of our avant-garde philosophy—
unexpectedly refreshing, a harmonious contrast."
Flavor Profile: **Delicate endive bitterness, sweet golden raisins, bright**
citrus. Unique, crisp texture. Sales & Marketing (Core Message)
Story Hook: "A daring frozen expression of our avant-garde philosophy—
unexpectedly refreshing, a harmonious contrast."
Flavor Profile: Delicate endive bitterness, sweet golden raisins, bright
citrus. Unique, crisp texture.

Ingredients (Key Components)
- **Pesto Base:** Fresh baby spinach, Belgian endive, golden raisins, toasted slivered almonds, garlic.
- **Flavor Boost:** Pecorino Romano (optional, for savory depth), EVOO, fresh citrus juice (e.g., Meyer Lemon).
- **Finish:** Golden raisin dust, fine citrus zest dust.

Citrus Strategy: Tailor acidity/aroma: Lemon (sharp), Meyer Lemon (sweet/floral), Blood Orange (vibrant).

Method (Core Steps)
- **Prep:** Blanch greens, ice bath, squeeze dry.
- **Blend:** Combine greens, raisins, almonds, garlic, Pecorino, EVOO, citrus. Blend smooth.
- **Freeze:** Pour mixture into shallow pans. Freeze solid.
- **Scrape:** Scrape frozen mixture with a fork every 30-60 mins until crystalline granita forms.
- **Serve:** Lightly mound granita. Finish with golden raisin dust and citrus zest dust.

Roasted Heirloom Carrot & Fermented Carrot Tartare with Green Pea "Caviar," Hazelnut Butter Crumble, and Carrot-Ginger Reduction

ELEGANT. INNOVATIVE. MASSIVELY PROFITABLE HAUTE CUISINE.

Concept: An incredibly elegant and innovative main, showcasing the humble carrot in multiple textures and flavor profiles, achieving richness through complex preparation and contrasting elements.

Why it works for Michelin/Vegetarian/Lactose-Free:

Heirloom Carrots: Offer natural sweetness, vibrant colors, and earthy depth. Roasting concentrates flavors.

Fermented Carrot Tartare: Adds a crucial layer of tangy umami and a unique texture, providing the "funk" often missing in vegetarian high-end dishes. (This utilizes a pre-prepared fermented carrot, or quickly 'pickled' carrot for home).

Green Pea "Caviar": A visually stunning, fresh, and slightly sweet element that mimics the luxury of caviar.

Hazelnut Butter Crumble: Provides a rich, savory, and nutty crumble, replacing cheesy gratins.

Carrot-Ginger Reduction: A concentrated, sweet-spicy, and glossy sauce that ties the dish together.

High-Margin Strategy: Carrot Tartare

This dish is designed for exceptionally high profit by leveraging low-cost core ingredients (carrots, peas) with high-value techniques (fermentation, spherification, precision reduction). The minimal fine-dining portion size further reduces per-plate COGS.

Cost Breakdown (Per Serving): Total COGS (Estimated): $1.73

Highest Cost Component: Hazelnut Butter Crumble ($0.94)– still a small fraction of the final price.

Low-Cost Core: Heirloom Carrots and Peas combine for less than $0.40 per plate.

Margin & Target Price: We target a Food Cost Percentage (FCP) of 20% for technical vegetable dishes.

Target Menu Price based on 20% FCP: $8.65 ($1.73 / 0.20).

Reality Check (Applying Fine Dining Multiplier): If the restaurant prices the dish at a realistic high-end main course price of $45.00:

Actual FCP: ($1.73 / $45.00) x 100 = 3.84%

This **exceptionally low FCP** demonstrates massive profitability. The WOW factor (skill, complexity, innovation) is the primary value driver, **justifying a menu price of $42 - $48**. The perceived luxury far outweighs the raw ingredient cost.

Roasted Heirloom Carrot & Fermented Carrot Tartare with Green Pea "Caviar," Hazelnut Butter Crumble, and Carrot-Ginger Reduction

MASTER RECIPE + FOH TRAINING

KEY INGREDIENTS (FOR 2):

- **Heirloom Carrots:** 6-8 medium, peeled. (Reserve some peel for optional garnish).
- **Fermented Carrots:** ½ cup, finely diced (store-bought or homemade lacto-fermented carrots; alternatively, finely diced carrots quickly marinated in apple cider vinegar, salt, and a pinch of sugar).
- **Green Peas (frozen):** ½ cup.
- **Agar-Agar Powder:** ¼ tsp.
- **Hazelnut Butter:** 2 tbsp (unsweetened, pure hazelnut butter).
- **Panko Breadcrumbs (gluten-free if needed):** 2 tbsp.
- **Fresh Ginger:** 1-inch piece, grated.
- **Orange Juice:** ¼ cup.
- **Vegetable Broth:** ½ cup.
- **Rice Vinegar:** 1 tbsp.
- **Maple Syrup/Agave:** 1 tsp.
- **Fresh Chives:** 1 tbsp, finely minced.
- **EVOO, Sea Salt, Black Pepper.**

METHOD: PREP & COMPONENTS (MISE EN PLACE)

Carrot Roasting:

1. Toss peeled heirloom carrots (6-8 medium) with EVOO, sea salt, black pepper, and fresh thyme sprigs.
2. Roast at 200°C (400°F) until tender and lightly caramelized.
3. Portion: Cut some into elegant lengths (e.g., 2-inch cylinders, bias cuts), dice remaining for tartare. Reserve some roast carrot peel, if desired, for crisp garnish.

Carrot-Ginger Gastrique (Reduction):

1. Sweat grated fresh ginger (1-inch piece) in EVOO.
2. Deglaze with orange juice (¼ cup).
3. Add vegetable broth (½ cup), rice vinegar (1 tbsp), and maple syrup/agave (1 tsp).
4. Simmer and reduce to a syrupy consistency. Strain through a chinois for a refined, clear gastrique. Hold warm.

Fermented Carrot Tartare:

1. Finely dice fermented carrots (½ cup). (If using alternative, quickly pickle diced carrots in hot apple cider vinegar, salt, pinch sugar, then chill and drain).

2. Combine with a small portion of diced roasted carrot for textural variation, finely minced fresh chives (1 tbsp), and a tiny drizzle of EVOO. Season lightly with salt/pepper.
3. Set aside for plating.

Pea Agar Spheres (Faux Caviar):
1. Blanch frozen green peas (½ cup) briefly, then shock in ice bath. Blend very smooth, strain through a fine mesh sieve. Reserve 2 tbsp pea liquid.
2. Heat the 2 tbsp pea liquid gently; whisk in agar-agar powder (¼ tsp) until fully dissolved and mixture just comes to a simmer.
3. Using a spherification dropper, carefully drop the hot pea-agar liquid into very cold, neutral oil (e.g., grapeseed or light olive oil) to form small spheres.
4. Collect spheres, rinse gently in cold water, and drain. Hold chilled.

Hazelnut Panko Crumble:
1. Combine hazelnut butter (2 tbsp), panko breadcrumbs (2 tbsp), a pinch of sea salt, and a tiny drizzle of maple syrup in a non-stick pan.
2. Cook over medium heat, stirring constantly, until panko is golden brown, toasted, and forms a fragrant, dry crumble. Cool.

Plating (Per Order)
1. Arrange elegant lengths of roasted carrot artfully on a small plate or tasting spoon.
2. Gently place a quenelle or small ring-molded portion of the fermented carrot tartare.
3. Scatter a few pea agar spheres around the plate. Spoon a delicate drizzle of the carrot-ginger gastrique.
4. Sprinkle the hazelnut panko crumble for texture. Garnish with micro-chives or a crisp of reserved carrot peel, if desired.

5. Season lightly with a final pinch of flake sea salt and fresh black pepper.

Roasted Carrot & Fermented Carrot Tartare- FRONT OF HOUSE TRAINING

The Story Hook (Innovation/PROFIT Factor):
"This is an example of vegetable-forward high-craft. We apply complex techniques—roasting, fermenting, reduction—to the simple carrot to create a luxurious and memorable centerpiece."

The Components: The plate features two distinct carrot flavors: Warm, earthy roasted heirloom carrots and a bright, tangy fermented carrot tartare.

The Innovation: The playful Green Pea 'Caviar' (gel spheres) provides a burst of freshness, perfectly balanced by the rich, savory crunch of the Hazelnut Crumble.

The Experience: It's a study in extremes: Sweet (reduction), Earthy (roasted carrot), Sour (tartare), and Crunchy (crumble). It feels substantial and luxurious.

Dietary Notes: Naturally Vegetarian and Gluten-Free. Contains Nuts (Hazelnut).

Sales Tip: Emphasize the technical complexity (fermentation, caviar) to justify the price and showcase the kitchen's skill.

FOH Upsell: *Enhance your dish with*:

- A whisper of White Truffle Oil or Aged Balsamic Glaze?
- Or a delicate sprinkle of Smoked Maldon Sea Salt?
- Our sommelier suggests a perfect Sake or Amuse-Bouche Cocktail pairing.
- For indulgence, consider a shaving of Cured Duck Breast."

Method: Lacto-Fermented Carrots
Yield: ~1.5 Liters | Prep Time: 30 min | Ferment Time: 5-7 days

Equipment: 2L Fermentation Crock/Jar with airlock, or sanitized jar with follower/weight and lid.

Ingredients:
- **1.2 kg (approx. 6-8 medium) Heirloom Carrots**, peeled, cut into uniform spears/discs
- **2% Brine Solution:** 1.5 L filtered water, 30g non-iodized sea salt
- **Optional: 1 tsp mustard seeds, 2-3 bay leaves, fresh dill sprigs (for flavor)**

Method:
1. **Sanitize:** Thoroughly clean and sanitize all equipment (crock/jar, weights, utensils).
2. **Brine:** Dissolve salt in filtered water to create a 2% brine.
3. **Pack:** Tightly pack uniform carrot pieces and optional aromatics into the sanitized crock/jar, leaving 2-3 cm headspace.
4. **Cover:** Pour brine over carrots, ensuring they are fully submerged. Use a fermentation weight/follower to keep carrots below the brine line.
5. **Ferment:** Seal with an airlock or cover tightly. Ferment at cool room temperature (18-22°C / 65-72°F) for 5-7 days.
6. **Monitor:** Bubbling indicates activity. Taste daily after day 4. Look for pleasant tang, slight fizz.

Quality Check: Brine should remain clear, no off-odors, no mold on submerged product.

Store: Once desired flavor is achieved, remove airlock/lid. Transfer carrots and brine to sealed containers. Refrigerate at 2-4°C (35-40°F) for up to 2-3 months.

Veridian: A Smoked Mustard Green & Spice Elixir
The Drink That Breathes in Smoke

In a world of fast comforts, **Veridian** insists on slowness. It begins with mustard greens — sharp, assertive, untamed — softened through the meditative patience of smoke. Half are transformed by fire and applewood, half by the dry kiss of the pan. Together, they create a foundation that is both bitter and warm, green and embered.

To this base, we add layers: the floral brightness of cardamom, the clean spark of white pepper, the grounding hush of cinnamon. Sweetness comes in whispers, not floods — date juice tempered by a simmered fig, balanced by the tang of tamarind. What emerges is a liquid meditation: smoke, verdancy, spice, and soul in quiet conversation.

"Veridian is not merely a beverage — it is transformation in a cup."

The Raison d'Être
At The Culinary Traveler, we search for the thresholds where opposites touch. *Veridian* lives at one of these crossings. Bitterness becomes warmth, smoke becomes clarity, spice becomes contemplation. It is not a drink designed to comfort in the usual way. It is meant to awaken. To remind you that everything sharp can be softened, and everything harsh can carry hidden sweetness.

The Recipe (Yields 4-6 servings)
Ingredients
- ⅔ cup mustard green leaves (young, robust)
- ⅓ cup fresh spinach leaves
- Applewood chips (or other smoking wood)
- 4 cups filtered water
- 8–10 green cardamom pods, lightly bruised
- 1 tsp white peppercorns, cracked
- 1 shard cinnamon bark (cassia or Ceylon)
- 2 tbsp date juice
- 1 dried fig, quartered
- 1–1.5 tbsp tamarind pulp (or 1 tbsp paste)
- Pinch of fine sea salt

Method
Smoke & Toast the Greens
1. Cold-smoke half the mustard greens and spinach for 20–30 minutes over applewood until fragrant.
2. Toast the remaining greens in a dry skillet until edges char.
Infuse the Spices
3. In a pot, simmer water with cardamom, white pepper, and cinnamon for 10–12 minutes.
Build the Base
4. Add the smoked and toasted greens. Simmer gently for 5–7 minutes.
Sweeten & Balance
5. Strain the liquid. Stir in date juice, the simmered fig (remove before serving if desired), tamarind, and a pinch of salt. Taste and adjust.

Serving Ritual
Ladle into a hand-thrown ceramic mug. Garnish with a smoked mustard leaf, or finish with a drizzle of tamarind-date reduction for a first sip of sweetness that yields to smoke.

Tasting Notes
Aroma: Green woodsmoke, sweet spice, faint resin.
Palate: Bitter edge softened into warmth; bright top notes of cardamom; white pepper flicker; deep green body; lingering tang.
Mood: Meditative, grounding, expansive.

Why Veridian Matters

Veridian reminds us that food and drink are more than sustenance – they are transformations. Just as greens yield their soul through smoke, so can we. This is a drink for winter evenings, for long conversations, for solitary moments when the world needs to slow down.

It is not crafted to please everyone. Instead, it is crafted to invite you inward. To pause, sip, and taste what it means for something bitter to become beautiful.

Savory Pear & Leek Cake Duet
THE FINAL BITE

The Strategic Advantage: 10/10 WOW and 10/10 PROFIT

This dish is a financial powerhouse because the WOW factor is entirely based on technique and concept, not expensive ingredients.

Savory Pear & Leek Cake Duet

Experience a symphony of unexpected flavors and textures: a warm, tender cake infused with deeply caramelized onion and sweet pear, served alongside a chilled, dark-crusted cannelé featuring a delicate leek and Mascarpone custard. Each component offers a distinct journey—from savory sweetness and subtle spice to a refreshing herbaceous finish—culminating in a truly unique and memorable culinary statement. Finished with ginger-poached pear, crème fraîche, and a toasted black sesame crumble, this dessert challenges tradition and delights the adventurous palate.

WOW Factor: Complexity & Surprise

- **Defiant Flavor Profile:** The unexpected introduction of caramelized onion and leek into a dessert elevates the dish from sweet to sophisticated. This savory-sweet balance is highly memorable and signals high-craft gastronomy.

- **Temperature Contrast:** The essential contrast of warm cake served next to the chilled Cannelé creates an immediate sensory experience that delights the guest.

- **Technical Excellence:** The inclusion of a perfect Cannelé—a famously difficult pastry to execute flawlessly—immediately justifies a premium price.

PROFIT Factor: Margin Mastery

- **Low-Cost Volume Base:** The cake is volumized by caramelized onions, one of the lowest-cost staples. This significantly reduces COGS compared to desserts reliant on expensive chocolate or nuts.

- **Batch Efficiency:** The Cannelé can be prepared and baked in large batches days in advance, eliminating complex on-demand labor and dramatically improving kitchen efficiency.

- **Portion Control:** Both the cake and Cannelé are formed in rigid molds, ensuring zero variance in portioning. This consistency protects the integrity of our sub-5% food cost target.

Savory Pear & Leek Cake Duet
METHOD + FOH SALES

This is the recipe for the main cake base, designed for precise portioning in ring molds.

I. Caramelized Onion Preparation (Umami Base)

1. **Ingredients:** 2 large yellow onions, thinly sliced; 2 Tbsp unsalted butter; pinch of salt.
2. **Method:** Melt butter over very low heat. Add onions and cook slowly for 45–60 minutes, stirring occasionally, until they are deep brown, jammy, and intensely sweet. Do not rush this step. Set aside to cool completely.

II. Cake Batter

1. **Ingredients:** 1 cup all-purpose flour; 1 tsp baking powder; 1/2 tsp salt; 1/2 cup (1 stick) brown butter, cooled; 3/4 cup packed brown sugar; 2 large eggs; 1/4 cup buttermilk (or whole milk); 1 tsp vanilla extract; 1/2 cup diced firm Bosc or Anjou pear.
2. **Method:** Whisk flour, baking powder, and salt together. In a separate bowl, cream the brown butter and brown sugar until light. Beat in the eggs one at a time, then mix in the vanilla.
3. **Combine:** Alternate adding the dry mixture and buttermilk to the wet ingredients, mixing only until just combined.
4. **Fold In:** Gently fold in the cooled Caramelized Onions and the diced pear.
5. **Bake:** Pour batter into prepared ring molds or a small cake pan. Bake at 350F (175C) until a tester comes out clean. Cool completely before cutting and warming for service.

Mascarpone & Sweet Leek Cannelé Recipe

This recipe is designed for a dense, savory-sweet custard center and a classic dark, caramelized crust.

I. Leek Infusion & Mascarpone Base

1. **Ingredients:** 1 cup whole milk; 1/2 cup heavy cream; 1/4 cup chopped sweet leeks (white/light green parts only); 2 Tbsp unsalted butter.
2. **Method:** Slowly sweat the leeks in butter until they are completely translucent and sweet—about 10 minutes. Do not brown.
3. **Infuse:** Combine the milk and cream, bring to a simmer, then add the sweated leeks. Remove from heat, cover, and let steep for 30 minutes. Strain the mixture, pressing gently on the solids.
4. **Base:** Whisk in 1/4 cup Mascarpone until smooth. Keep warm.

III. Cannelé Batter

1. **Ingredients:** 1/2 cup granulated sugar; 2 large eggs plus 1 egg yolk; 1/4 cup all-purpose flour; 1 Tbsp dark rum (optional, for flavor); 1 tsp vanilla extract.
2. **Method:** Whisk the sugar, whole eggs, and egg yolk until well combined. Gradually whisk in the flour until smooth. Slowly pour the warm leek/Mascarpone mixture into the egg mixture, stirring constantly.
3. **Rest:** Stir in the rum and vanilla. Strain the batter one final time for smoothness. Cover and chill for a minimum of 24 hours. This rest is mandatory for the final texture.
4. **Bake:** Pour chilled batter into pre-buttered and floured (or specialty Cannelé-waxed) copper molds. Bake at a high temperature 450F / 230C for 15 minutes, then reduce the heat 375F /190C and continue baking until the

exterior is dark mahogany brown and the interior is soft. Serve chilled.

Plating Components & Assembly

I. Ginger-Poached Pears

1. **Ingredients:** 1 firm pear, thinly sliced; 1 cup water; 1/2 cup sugar; 1 inch fresh ginger, sliced.
2. **Method:** Combine water, sugar, and ginger. Bring to a simmer until sugar dissolves. Add pear slices and simmer gently for 5 minutes until tender-crisp. Cool the pears in the syrup.

II. Toasted Black Sesame Crumble

1. **Ingredients:** 1/4 cup black sesame seeds, toasted; 1/4 cup AP flour; 1/4 cup brown sugar; 2 Tbsp brown butter.
2. **Method:** Pulse all ingredients together until a coarse, sandy mixture forms. Bake at 325F/160C for 10–12 minutes until crunchy. Cool and store in an airtight container.

III. Final Assembly (Side-by-Side Plating)

1. **Warm:** Gently warm one precise slice of the Pear & Caramelized Onion Cake.
2. **Place:** Place the warm cake slice on one side of the plate.
3. **Contrast:** Place one chilled Mascarpone & Sweet Leek Cannelé immediately BESIDE the cake. Ensure they do not touch to maintain the critical temperature contrast.
4. **Garnish:** Lean a slice of the Ginger-Poached Pear against the cake. Scatter a small pinch of the Black Sesame Crumble across the plate and around the components. add a very light whisk of the Crème Fraîche
5. **Serve immediately.**

Front of House Training (The Sales Pitch)

The Opening Line:

"Chef wanted to challenge the idea of dessert. This is the Pear and Caramelized Onion Cake—a masterpiece of savory-sweet balance."

Key Selling Points:

The Flavor Fusion: "The onion is caramelized for nearly an hour to unlock a deep, jammy, savory sweetness. It's balanced by the pear and the subtle, herbal depth of the sweet leek in the cannelé."

The Structural Contrast: "The dish delivers an immediate textural and temperature shift. You get the warm, moist cake served right next to the chilled, firm cannelé, whose dark crust provides a satisfying crunch."

The Final Hook: "This is truly a culinary statement—it's the dessert for the guest who wants something they have never experienced before."

Critical Server Notes:

Allergies: Contains Gluten, Dairy (Mascarpone, Butter), and Eggs. It cannot be modified.

Plating: Explicitly state the components are served side-by-side to avoid confusion; emphasize the temperature difference.

The Final Word: Transforming Craft into Cash Flow

This issue of The Culinary Traveler Volume 17 has provided the BLUEPRINT FOR MARGIN MASTERY, demonstrating how to Drive maximum cash flow in gastronomy by merging culinary prestige with unmatched cost efficiency. Our core MANDATE is fulfilled by delivering high-impact, scalable solutions where the WOW factor is generated by technique and concept, not by expensive ingredients.

From the Hyper-Margin utilization of negligible-cost leek ash in the Welsh Leek Ash & Black Garlic Umami Broth to the Defiant Flavor achieved by leveraging low-cost staples like caramelized onions in the Savory Pear & Leek Cake Duet, every featured recipe is engineered for massive profitability.

We have shown that true distinction lies in having complete control over flavor agents, whether through house-made foundations like the UK Heritage Luxe Kit (The Welsh Leek Ash & Black Garlic Umami Broth, The Mermaid's Larder Finishing Oil, The Captain's Brown Sauce) or by transforming classic items into 90%+ margin anchors, such as the Black Vinegar Empire Kit's high-wow, high-profit solutions like the "Fizz Phantom Cubes". The journey through Michelin Economy solutions, like the $1.73 COGS Carrot Tartare, confirms our approach of chasing the big WOW factor and the big Profit Margin, at the same time.

To successfully implement and scale these high-margin innovations in your operation—from the Endive & Spinach Linguine (90% Margin) to the highly complex, yet profitable Savory Pear & Leek Cake Duet (sub-5% food cost target)—you require the full strategic package.

Unlock Maximum Profit: The Complete Business Blueprint Kit

The strategies and concepts explored in this volume are ready for immediate deployment. For those seeking to turn these recipes into structured, high-volume profit centers, the full Profit Business Blueprint Kit is the essential, comprehensive tool.

For the complete Profit Business Blueprint Kits, including master recipe files, kitchen cards, scale production methods, cost/margin spreadsheets, and marketing and sales assets, visit:

www.explorechartatlas.com/blueprints